For centuries, people of the North
have used the resources wisely.
... Diavik is continuing this tradition.

Diavik

Our Foundation, Our Future

The Story of the Diavik Diamonds Project

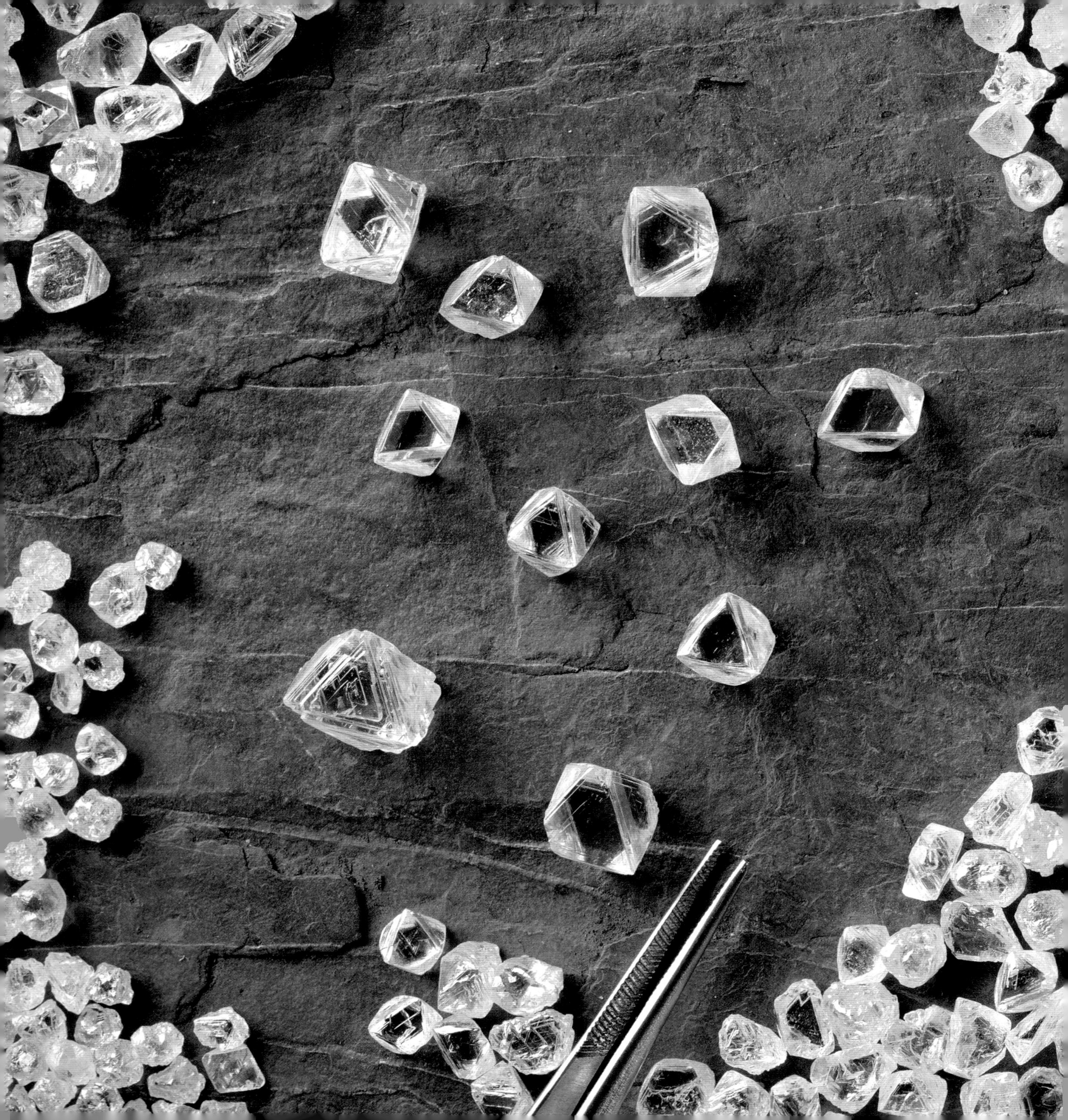

Vision Statement

Diavik's vision is to be Canada's premier diamond producer, creating a legacy of responsible safety, environmental and employee development practice and enduring community benefit.

Our Values and Principles

We treat ourselves and all those with whom we meet and work with dignity, consideration and respect.

We are committed to active partnership with local communities, businesses, governments and non-governmental organizations.

We are dedicated to meeting the needs of our customers and to contributing long term value to our investors and to the North.

We strive for excellence in all we do.

Diavik – Our Foundation, Our Future
The Story of the Diavik Diamonds Project

Editors: Laura E. Tupper, Sara E. Neamtz, GPC International
Design: Dot van Vliet, Outcrop Communications Ltd.

Cover: A 0.91 carat emerald-cut Diavik diamond reflects the beauty of the North.
Photo by Frans Jan Fortunati.
Page i: Lac de Gras

National Library of Canada Cataloguing in Publication Data

Main entry under title:
Diavik, our foundation, our future

ISBN 0-919315-32-1
1. Diavik Diamonds Project. 2. Diamond mines and mining – Northwest Territories.
I. Tupper, Laura E., 1956- II. Neamtz, Sara E., 1970-

TN994.C3D52 2002 338.7'622382'09719 C2002-910009-7

Printed and bound in Canada by Quality Colour Group, Edmonton.

Contents

Foreword

Northern peoples have lived for thousands of years in the place dubbed the "Barren Lands". They thrive by looking beyond the obvious; by being attuned to the bounty that the North can provide.

We at Aber Diamond Corporation take great pride in our founding role in the Diavik Diamonds Project. A decade ago, there were very few people who believed that northern lakes could hold the secret of diamonds beneath them – few could foresee the Diavik story as it has evolved. Patience, diligence and the efforts of resource experts and investors have been rewarded as events of the past decade have unfolded.

First, we had the good fortune of following in the footsteps of the discovery of Canada's first commercial diamond deposits made by Chuck Fipke and his partner Stewart Blusson. Their extraordinary feat of persistent exploration, over many difficult years, was the spark that ignited the diamond exploration boom with ourselves as participants.

Our next fortunate decision was aligning ourselves with Rio Tinto who have continuously provided steadfast technical and financial support since those early years. Following the discovery of economic deposits in 1994 and 1995, our joint venture arrangements with Rio Tinto were assigned to its subsidiary, Diavik Diamond Mines Inc.(DDMI), who now manages the project on behalf of us both.

By looking beyond the obvious, and true to DDMI's vision and principles fully supported by Aber, the project has been strengthened through the active participation of local communities and a shared respect for the land on which they depend.

That is why this book, the story of the Diavik Diamonds Project, starts where it should: with the story of the land and its people.

Robert A. Gannicott
President and Chief Executive Officer
Aber Diamond Corporation

Sunrise on early Diavik Camp

Introduction

Over one-third of the Northwest Territories is covered by lakes and rivers that are replenished each year by the seasonal melt of the winter snows, augmented by the occasional rainfalls in warmer months of the year. In spite of its apparent abundance, water is considered a precious resource, especially by aboriginal peoples. Not only has fresh water sustained generations, it has also provided habitat for much of the wildlife that is critical to their traditional lifestyles. These rivers and lakes have been the main transportation routes for centuries, allowing movement over what is otherwise difficult terrain. Communities have grown on the banks of these great waters which served to connect the people throughout the North.

The Diavik Diamonds Project has made its home in this land. And just as the waters of the North have come together to shape the lives of all who have the will and determination to make the North their home, a number of interests have come together to shape the Diavik Diamonds Project. The deep understanding and respect for the land of aboriginal people; the strong spirit and perseverance of all Northerners who search for opportunities for themselves and their children; the technical and scientific skills of the mining industry; and the financial resources of investors from around the globe, have come together to ensure the legacy of the Diavik Diamonds Project is one that supports the well-being of northern communities and the land, waters and wildlife on which they depend.

This is the story of the first years of Diavik, told through the eyes of the men and women who have been part of this great adventure – an adventure of discovery of mineral wealth, traditional knowledge and human capacity.

Stephen Prest
President
Diavik Diamond Mines Inc.

The crunch of hard packed snow.
The sting of a blinding blizzard.
The rush of a freshet during the melt.
The patter of rain drops.
The fresh breath of a heavy dew.
The crash of ocean waves.
The water of life, flowing tirelessly,
clean and pure.
From the sky, to the land, to the many
lakes and mighty rivers.
Transformed along the way as it
replenishes the land and all its creatures.
A gift from one generation to the next.
A treasure to be guarded for eternity.

Headwaters of the Coppermine River

CHAPTER 1

The Natural Endowment

The people of the Northwest Territories are people of the land. They have always depended on the resources that come from the land – the animals, the water, the plants, the herbs.

Above: Red-necked phalarope

Left: Bearberry in the fall

At the height of summer, berries beckon from every side, nets are loaded down with fish, and wild game abounds. But in the depths of a four-week, forty below zero stretch of winter, the land can appear terrifyingly harsh and austere. The Dene and Inuit, who have lived on the land for untold generations, have tales of dark times and starvation.

Despite its sometimes forbidding aspects, Canada's North has attracted, and continues to attract, people who dream of adventure, wealth or simply a different way of life. After Alexander Mackenzie descended the great river now bearing his name to the Arctic Ocean, a legion of fur traders followed in his wake.

The North also has a long tradition of mining. In 1578, English adventurer Martin Frobisher took tons of rock back to England from Baffin Island thinking it contained gold. It did not, but that early disappointment did not discourage those that followed.

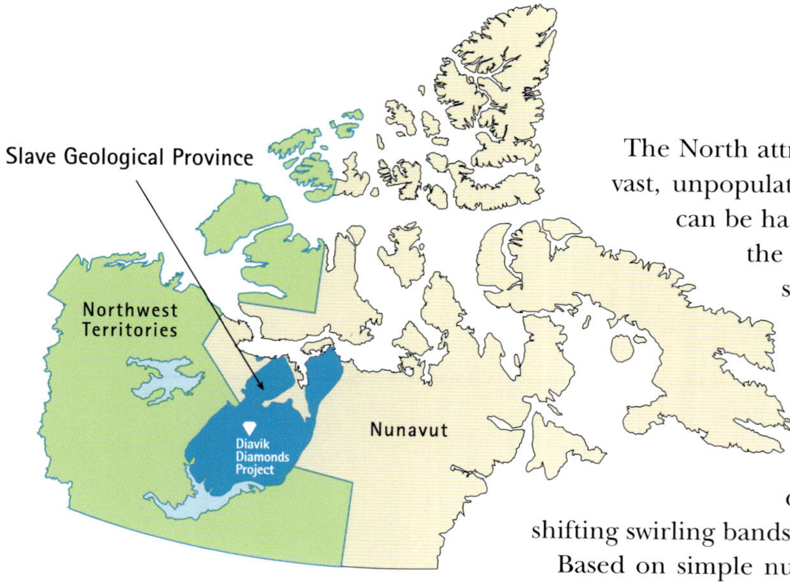

The North attracts people who appreciate its rugged beauty and vast, unpopulated spaces. While it is true that the environment can be harsh and inhospitable, it is, at the same time one of the most beautiful places on earth. Winters are long; snow and ice cover the ground for eight months of the year and daylight can be uncomfortably scarce. But what sun there is usually illuminates brilliantly clear days, the crust of the packed snowdrifts sparkling like billions of diamonds. In the hours of darkness, the sky is often lit up by the dazzling natural display of the Northern Lights, shifting swirling bands of vivid colour.

Based on simple numbers, Canada's North which includes Nunavut, Yukon and the Northwest Territories, is a relatively empty space at the top of the world. The Northwest Territories, home to Diavik, has a population of around 40,000. With an area of approximately 1.2 million square kilometers, this gives it one of the lowest population densities in the world, with less than two people per hundred square kilometres. The capital, Yellowknife, is the largest community in the Northwest Territories with a population of about 17,500.

The unique nature of the land has created resilient, innovative people. More than half of the North's population is of Dene, Inuit and Métis descent. Aboriginal beliefs and customs are strongly embodied in northern society. This includes a deep and abiding respect for the power, beauty, and integrity of the land.

Ancient Rocks of the North

The Northwest Territories is split into five geological provinces: the Slave, the Churchill, the Bear, the Cordillera, and the Arctic/Interior platforms. At over 2.5 billion years old, the rocks of the Northwest Territories are some of the oldest in the world. This ancient rock contains important mineral deposits including gold, copper, zinc, nickel, and yes, even diamonds.

Geography and Environment

Much of the Northwest Territories is within the basin of Canada's longest river, the Mackenzie, or "Deh Cho" (big river) as it is known in the language of the Dene who live along its course. From Great Slave Lake (the tenth largest lake in the world), it flows 1,740 kilometres north, collects the outflow of the Great Bear Lake (the eighth largest lake in the world), and finally empties into the Beaufort Sea on the Arctic coast. Unlike many other northern rivers it has no major rapids or waterfalls, which made it an important navigation route for aboriginal people, early explorers, traders

The Northern Lights

The spectacular beauty and magnificence of the natural phenomenon known as the northern lights or Aurora Borealis are reflected in traditional northern culture. Their mystical dance has captured the imagination of generations.

The oral history of the North abounds with explanations of the spellbinding northern lights. Some believe that the Aurora Borealis are the spirits of deceased ancestors or an entry way to the spiritual world. Others believe that the lights are a connection to the creator and that through them, one can embark on a spiritual journey. It is often said that the caribou are gathered underneath the northern lights and that their magical brilliance is a reflection of the ice that is cleared by the scratching of the caribou's hooves.

Other cultures have linked the lights to the fruitfulness of the earth, a torch-lit procession or a joyful dance.

The enduring and unique qualities of the northern lights have been adopted by DDMI in its corporate logo to reflect the commitments it has made to the people of the North.

"I have no ambivalent feelings about the North. I've loved it and hated it at times, but I never thought it was just okay. It's either doing its worst to you or you're standing in the most beautiful place on the planet. It's hard to explain to people, you have to experience it."

Ian Graham
Geologist on the Diavik Diamonds Project

and prospectors. Another major river in the Northwest Territories is the Coppermine, which flows north to the Arctic Ocean from Lac de Gras, where Diavik is located.

Apart from the Mackenzie Mountains on the western edge, the Northwest Territories is generally flat, glaciers having scraped much of the rock clean. There is water almost everywhere: lakes, ponds, streams, and soggy muskeg swamps. The "taiga", or boreal forest, covers the land below the tree line. Spindly pines, spruce, birch, and aspen are the most common species. To the far north and east of the territory is the arctic tundra region. It is too cold for trees here: low-growing shrubs and grasses cover the land in the brief northern summer.

Red fox

Canada's North is also home to numerous large and small mammals, countless birds, fish, and insects. Many of the birds are migratory: one-fifth of the continent's ducks, geese and swans summer in the North. The caribou are also migratory, although both their winter and summer homes are in the North. For the winter, they move down to the tree line, where temperatures are relatively warmer. In spring, they move back out to the tundra, where the rich fodder will fatten their newborn calves. Caribou herds can number in the hundreds of thousands. They were traditionally a principal source of food, clothing and shelter for many of the North's aboriginal peoples. Today, caribou meat remains a staple in the diets of many aboriginal people and is their main source of protein. The caribou also play a key role in aboriginal culture and spirituality. Caribou leather covers the drums played by Dene peoples in times of prayer and celebration. Other northern animals are also locally important for food, pelts, or increasingly, for tourist attractions. Moose, musk ox, bears, bison, fox and wolves are found in various parts of the Northwest Territories.

CHAPTER 2

Stewards of the Land

"[There are] things unseen, the sounds of animals, the wolf's cry, echoes wailing around us. We hear these things unseen. These were believed. Now we only believe what we see, not what we don't see. That is why things have been changed. The spiritual world is unseen. This comes to us from our ancestors. The spiritual world has to be made clear to our children. We have to make our children believers again. They have to learn to interpret the unseen ..."

Elizabeth Mackenzie, Yellowknife, June 1991

Above: Timber wolf

Left: Dogrib elder Nick Black was a skilled canoe-maker. Using traditional knowledge, he combined birch bark, pine, and spruce gum into light-weight, functional works of art. As a symbol of the value of traditional knowledge across generations and cultures, Diavik commissioned two of Nick Black's canoes.

The Dene and the Inuit have made their home in Canada's North for millennia – long before the arrival of European explorers.

In addition to the Gwich'in, there are several Dene tribal groups in the Northwest Territories, including the Dogrib (Tlicho), the Yellowknives (T'atsaot'ine), the Chipewyan (Denesuline), South Slavey (Deh Gah Got'ine) and North Slavey (Sahtu Gah Got'ine).

The Dene were great hunters who traveled in family groups throughout the North, dog sledding in winter and canoeing in summer through the numerous rivers and lakes. Hunting caribou was vital to the Dene existence.

Camp size and movement were based on seasonal availability of wildlife and fish, especially the caribou migration. It was considered a rite of passage into manhood when a boy made his first caribou kill. Caribou were usually taken using snares set along driving fences or spearing as the animals crossed rivers and lakes.

The Inuit inhabit the far North, from Alaska to northern Quebec and Labrador. Their beliefs were also based on respect of the land and its resources. They hunted land animals but they also took advantage of the abundant population of sea mammals and fish along the arctic coast. The Copper Inuit harvested arctic char at the mouth of the Coppermine River near today's community of Kugluktuk.

Fur traders were the first reported Europeans to reach the Great Slave Lake area. In 1769 Samuel Hearne came in search of copper. Guided by

Chipewyan Chief Matonnabee, Hearne reached the Coppermine River in 1772, encountering both the Yellowknives and the Dogribs in his travels.

Over time, more Europeans came and settled in the North. They established trading posts, missions and small communities. They relied heavily on the aboriginal population as guides and for basic survival. Some people of European descent, particularly the French-speaking voyageurs in the fur trade, married and had children with aboriginal women. The French called the descendants of these couples Métis. The Métis were often important in dealings between First Nations and newer arrivals, acting as interpreters and middlemen. The Métis formed their own communities and developed their own language and culture.

Traditional Knowledge

Prior to the written word, children would listen as elders spoke about their lives and of the ways the people lived on the land. As they got older, they would learn by doing, guided by elders and other experienced members of their community. Generation after generation passed on knowledge, legends and traditions in this manner. This traditional knowledge is still being passed on, both orally, and in written and recorded forms.

Richard Weishaupt draws on the experience of a group of elders.

A Time of Change

Within the broader political context, change is occurring in the communities of the Northwest Territories. In northern aboriginal communities, there is a strong belief that change must be motivated and endorsed by the grassroots, and only by improving conditions in the communities will opportunities be created for self-sufficiency.

Dogrib Grand Chief Joe Rabesca, commenting on the future, stated, "The future of our people will depend on the development of our resources and lands. We can no longer rely on trapping as a way of supporting our families and communities. Very few people trap because of many reasons including the low fur prices and the high cost of going out on the land. More and more young people are becoming better educated as time goes on, and we can't expect them to have a good living off trapping. It just can't be done anymore. What we need is to create more jobs that our young people can fill – jobs that are long term. We know these jobs cannot be created just by what's happening in our communities. Jobs will have to be created by other activities, such as mining and other things that are outside our homes. This, we know for sure. We are interested in construction, trucking businesses, petroleum distribution enterprises, facility management, catering and aviation, all of which are related to resource development. We need to focus on developments which have the opportunity to create long term wealth for our people."

The aboriginal peoples of the North have survived through the centuries based on their intimate knowledge of and respect for the land and its resources. They believe that their continued survival depends on them being respectful in their relationship with the land, its animals, the spiritual world, and other people.

DDMI respected the wealth of knowledge possessed by aboriginal elders. DDMI consulted them extensively to gain an understanding of how best to protect the land.

"Very early in the project, Diavik recognized that traditional knowledge was valuable and we sought opportunities for aboriginal elders and others in the communities to explain it to us," said Doug Willy, former Vice President, Community and Government Relations. "It was then our responsibility to ensure that it was incorporated in the design of the project."

DDMI approached elders in the communities and asked for input on how to design the mine site. Richard Weishaupt, former Manager, Northern Affairs, said, "Early in the project, we brought some elders up to the site and asked them what they thought of where the project team were thinking of putting the camp. The elders said we'd have snow drifting constantly into the camp, but if we put it in another spot the snow drifting would be less. And we listened."

The elders' suggestion about the placement of the camp and many other recommendations arising from the company's consultations with northern and aboriginal communities have been implemented in the permanent project design.

Gord MacDonald learns more about the land and water near the Diavik site from Peter Kamingoak and Joe Allen.

CHAPTER 3

Diamonds in the North

"As soon as people started talking about diamonds, everyone wanted part of it. We worked closely with Diavik to find ways to become involved in the industry, not only through employment but also by starting up competitive businesses to supply the mine. At the same time, we realized that in 30 or 40 years these mines will not be here. The investments that we are making now, like training and education, are needed to support our communities in the future. The real benefits from the diamond mines are not for me or my children, they are for the future generations."

Ted Blondin
Dogrib Treaty 11

Despite a mining history dating back to the 1920s few people suspected that there were diamonds in Canada's North. Pictured are the legendary bush pilot, Punch Dickins, and prospector, Gilbert Labine.

Left: Temporary camps, such as this one at the southeast end of Lac de Gras, were home for the Kennecott exploration team.

Diavik

Many people are unsure of the origin of the word Diavik. The word Diavik was created by Lee Barker when he combined the first letters of the word "diamond" and the letters "Vik", which came from West Viking Exploration, one of the original companies involved in the Lac de Gras discoveries.

At the turn of the 1990s, many people reacted with considerable skepticism on hearing the news of the widening search for diamonds in Canada's Northwest Territories. Some diamond-bearing rock had previously been discovered in Canada's North, but the diamonds were small and discoloured, not of any economic value. Most geologists had written off the area as a diamond prospect. Few people expected to ever find worthwhile diamond deposits in the North.

The 1991 announcement that diamonds had been discovered under a small lake just north of Lac de Gras set off one of the biggest mineral prospecting rushes ever in Canada's history.

A small private company called West Viking Exploration, managed by Lee Barker, was one of the first to respond. Barker had many contacts throughout Canada and the Northwest Territories and quickly assembled a team with diamond expertise. Grenville Thomas, a West Viking Director with a career-long association in Canada's Northwest Territories, promptly engaged the help of Covello, Bryan & Associates, a local Yellowknife exploration service company to assist West Vikings' prospecting activities. Shortly after, the key areas were secured where the Diavik diamonds were subsequently found. Then West Viking merged into Aber Resources Ltd., with Barker continuing as President during the early years of the Diavik Project.

Gren Thomas, Corporate Director, Honourary Chairman of Aber Diamond Corp.

Andrew Gorlay, Karl Chappel, Mike Tooley, Richard Weishaupt, Tobey Hughes, Dave Kelsch and Leilagh Tate, members of the Kennecott exploration team.

Kennecott Canada Inc. and the Joint Venture

One of the strengths of the junior mining companies is their ability to identify and pursue a good opportunity. Building on their initial success often requires substantial capital investments. Recognizing the potential of the Aber prospects, Kennecott Canada Inc. (KCI), a subsidiary of Rio Tinto plc (Rio Tinto), entered into an exploration arrangement with Aber.

"At this point in time, there was very little in the way of scientific approach to determining what might be the most prospective areas," said former KCI President John Stephenson. "Our strategy was simply to acquire a significant interest as close to the initial discovery area as possible. The principle of "closeology" was adopted. We had no idea which claims, if any, might contain diamondiferous kimberlites. But we were hopeful the Aber claims would pan out."

Diavik uses helicopters to move exploration drilling equipment over the sensitive tundra.

Above: Jennifer Burgess examines a kimberlite sample.

Right: The Aber Diamond

The Aber Diamond

Once the joint exploration arrangement was signed between Aber and KCI, work on Diavik progressed quickly under the management of Bob Hindson, Aber's Vice President Exploration. Stephenson, along with Buddy Doyle, Exploration Manager, Diamonds, at Kennecott Canada Exploration Inc., directed the project. The Diavik team soon found more than 30 kimberlite pipes, but none had any significant quantities of diamonds. The first major discovery was in March 1993, with a pipe called A21. "We can tell if these indicator minerals are hot or cold and the ones for A21 were hot," recalls Buddy Doyle. "We knew there was something out there. So we did more work on A21 and it showed up the next year."

Just before the ice started to break up in 1994, they began to work on another target, A154, which had been selected by Joe Inman, KCI's Chief

Geophysicist. After ten days of drilling, the drill cores showed no sign of kimberlite. They were only budgeted for one hole and it was late in the season.

The team decided to gamble on one last hole before spring break-up and moved the drill 15 metres to the south. "Finally, we got a discovery hole down where we struck kimberlite, but then we lost the ice and had to pull the rig off," said Eira Thomas, the Aber Project Geologist. "So we finally had the discovery hole core back in the core shack. We waited until almost everybody had turned in for the night before we unboxed the core."

Robin Hopkins, another Aber geologist, broke the core in half on a natural break. On one side he saw a large indentation. On the other half of the core was a bright, crystal-like stone.

That moment is etched in Eira Thomas' memory, "Robin holds up the core and said, 'Hey what's this?' And we all looked over, and there it was. This two and a half-carat diamond sticking out of the core. It was unbelievable."

Diamond drillers retrieve rock core samples from far beneath the surface to help reveal hidden ore deposits.

Buddy Doyle remembers his reaction, "Sure I knew it was a diamond, but I couldn't believe it. I have friends who have worked in diamonds for over 20 years, and I knew they had never seen a diamond like this one. So I thought, 'maybe this isn't a diamond.' And I started flicking to see if it was something else. But it wasn't. It was a diamond."

That night, after a brief celebration, Eira Thomas slept fitfully, with the diamond core under her pillow. The next day she took the core to Vancouver to show her dad, Gren Thomas. She insisted he come to the airport to pick her up, and Thomas grudgingly agreed. When she showed him the diamond, Thomas' eyes popped out of his head. He could not believe it.

"Is this for real?" he asked. His daughter smiled and replied, "Yes. It's real."

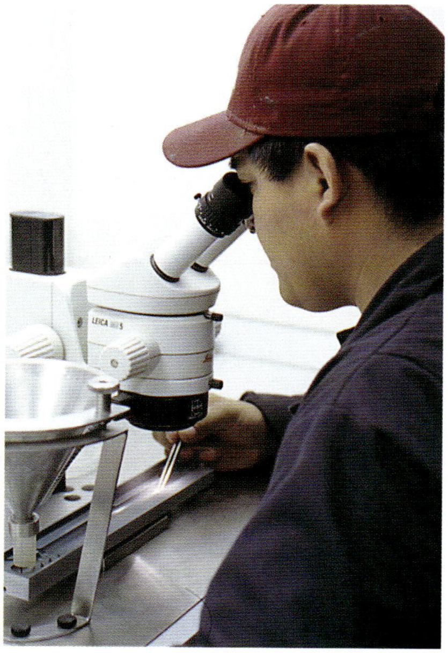

Steve Itsi examines the bulk sample under a microscope

A Project Confirmed

Even though the discovery of the Aber Diamond was significant, Aber and KCI still had to prove the kimberlite pipes that they had found were valuable enough to mine. The Aber Diamond was exactly the evidence the team needed to justify additional funding to move to the next stage of the project.

"The discovery of the Aber Diamond was a turning point for the project," said John Stephenson. "It quickly became evident that the level of funding needed to be increased to not only retrieve a bulk sample from this pipe, but to search for and evaluate other undiscovered pipes that might be present and close by."

It turned out that A154 actually contains two pipes and that explained why the team had missed when they first drilled into the A154 target – they had drilled between the two pipes. In March 1995, Dave Kelsh, former KCI Project Geologist, and his team discovered a fourth kimberlite pipe, A418, that had been missed in the original airborne surveys but was right beside the A154 pipes.

The next step was bulk sampling, in which large samples of kimberlite from A154 South and A418 were mined from underground workings situated under Lac de Gras, and accessed from a mine shaft entrance on East Island. Richard Lock, a mining engineer with experience in diamond mines in South Africa, managed the operation. "This was the first time I had ever mined with a large lake only a few hundred feet above me. On top of this, we were trying to maintain a schedule that would allow us to ship the kimberlite to Yellowknife on the winter road. It was cold and wet. I couldn't have been further away from the sunny skies I had left behind in South Africa."

The kimberlite ore that was recovered from bulk sampling was processed at the Diavik pilot plant in Yellowknife. In 1997, Harry Ryans, Manager Process Engineering, added extensive diamond processing experience he gained while in South Africa to the DDMI team. Except for Ryans, all of the people operating the processing plant were Northerners.

The Diavik Project site during bulk sampling. The underground workings required to retrieve the bulk samples of kimberlite ore were accessed by a large steel "culvert-like" entrance.

The conclusion of the studies on the Diavik pilot plant diamonds during 1996 through 1998 warranted pursuing the next stages of the project.

"The results of the bulk sampling were really quite positive. We knew that in Diavik we had diamonds that could meet the high quality standards demanded by top diamond retailers around the world" recalls Andy Davy, former DDMI Project Consultant.

Gord Hosick, Hay River, underground with Redpath miner during bulk sampling.

Quality diamonds

Of the 21,000 carats of diamonds recovered in the Diavik pilot plant, fully one-third of them are considered gem quality. Those diamonds were sent to Rio Tinto's Argyle Diamonds in Australia, where they were cleaned and sorted. Some of the high-quality, larger diamonds were cut and polished.

CHAPTER 4

Setting the Course

"It helped Diavik quite a bit because they hired people from the North. You know, there was Doug Willy, Tom Hoefer, Doreen Sangris, Erik Madsen, Eric Christensen and a lot of other people that we knew. We saw that as important."

Darrell Beaulieu, CEO, Deton'Cho Corporation

Above: Yellowknife, the "Diamond Capital of North America", reflects the contrasts of the North.

Left: Standing on the original runway, Rod Davey discusses the future plans for developing the Diavik site with representatives from the Kitikmeot Inuit Association.

In the early days, the project activities were managed by Aber and then subsequently in 1995 by Kennecott Canada. In November 1996, as part of a corporate reorganization, Diavik Diamond Mines Inc., (DDMI), was created to manage the Diavik Project. Even though the mining industry had been in the North for almost a century, DDMI became the first major mining company to establish its head office in Yellowknife, the "Diamond Capital of North America".

Rod Davey, a long-standing Kennecott executive, was brought on to the Diavik team in 1995 and became the first President of DDMI. Diavik was more than a mine to Davey – it was an opportunity to build something that would have lasting and positive effects on communities and people.

One of the first people hired in late 1995 in anticipation of the mine going forward was Doug Willy. He had many years of experience working

"Early on, DDMI encouraged northern people to participate in training provided by the company in fields related to mining. For example, fifteen interested young people from local communities were trained in the field by qualified geologists and engineers to assist in a number of geotechnical duties. Every one of these trainees received a certificate from the program with most securing related jobs that winter."

Tom Hoefer
Manager Public and Government Affairs

John Wifladt was one of the first Diavik trainees.

in the northern mining industry. Between 1972 and 1980 Willy and his family lived and worked in a number of small aboriginal communities throughout the North. Despite his Saskatchewan roots, in his heart, Willy was a Northerner. Based on his life experience, he brought to DDMI an appreciation of the Dene and Inuit peoples and cultures. The combination of the way Rio Tinto works, Willy's background and Davey's leadership set the stage for the active participation of the local communities in all aspects of the project.

People and Plans

The decision was made to set up shop in the North – "It was an easy decision" said Davey, "because it was the right one."

DDMI committed itself to providing northern employment during the construction, operation and eventual closure of the mine. As a goal, DDMI's aim is to reflect the demographics of the region in its workforce, with priority being given to neighbouring communities in the Northwest Territories and Nunavut. DDMI's contribution to capacity building in the North includes community-based and on-the-job training programs, trade apprenticeships, scholarships and northern business development.

DDMI's training programs have been supported by a number of partners including, among others, our participating communities, local businesses, the territorial and federal governments and Aurora College. According to Maurice Evans, President of Aurora College, "Diavik's dedication to meeting their northern hiring targets has made them one of the most effective training partners that we have been involved with. Aurora College and Diavik have been innovative and solution-oriented when faced with training challenges. We use the College's expertise with northern training, our existing infrastructure and new facilities provided at the Diavik site to deliver programs from skills-based training to introductory trades. The community-based approach of some of these programs has also been a real plus for the North. Common goals and trust are the keys to the success of this partnership."

Northern business development was another aspect of capacity building that DDMI focused on. In 1996, DDMI approached the Yellowknives Dene to explore setting up a joint venture relationship so its community

Eka'ti Services supplied Diavik's pre-construction camp at the site.

could benefit from an ongoing business relationship with the project. Eka'ti Services was formed by the Yellowknife Deton'Cho Corporation and entered into a contract to supply catering and food services to the Diavik camp, first for the exploration, and later for the construction of the mine.

As project activities intensified, Eric Christensen, Venture Development Manager, created the Northern Business Opportunities Profile, which divided the project into construction, mining and reclamation phases and showed where local businesses as well as individuals could capture the opportunities created by the Diavik Project.

"We can only achieve our potential in the North and elsewhere if we expect all of our citizens to do well in education, work and life and if we willingly support them in that achievement. We are already seeing progress. Many of our young people are very technologically savvy, outward looking and more educated than ever. They are managing very well. However, we are more determined to build on this momentum."

Hon. Ethel Blondin-Andrew, M.P., P.C.
Member of Parliament, Western Arctic

CHAPTER 5 A Time to Listen

"At a meeting in Kugluktuk, I distinctly remember Joe Allen [Evyagotailak], a prominent resident, telling us that the Coppermine River is the pipeline that carries water to his people. Lac de Gras is their water tank. I never heard anybody explain it the way he did, but it made a lot of sense. And here we were building a mine in their water tank. We had to ensure that we protected their water tank."

Erik Madsen, DDMI Environmental Manager

Above: Cloudberry

Left: Isadore Sangris, Steve Nitah, Michel Paper and Erik Madsen exchange ideas at the Diavik site.

After the discovery of the Aber Diamond in 1994, Buddy Doyle remembers those early discussions as to what would be the right way forward for the project, "As soon as I saw that diamond I knew this project had potential and there was going to be a mine. One of the biggest challenges facing both us and our neighbouring communities was to ensure the company addressed environmental issues and potential impacts on the communities."

If Doyle was right and there was going to be a mine on East Island, the company would have to study all the factors in the landscape – the vegetation, the animals, the water, and the air quality. It would also have to take into account the socio-economic status and needs of local people.

Mike Baker conducting field studies

"We collected a wide range of baseline material, from caribou to bugs, you name it – traditional knowledge, people, socio-economic aspects ... We got all that information that the engineers used while developing this project. We reviewed design alternatives with the communities to seek their advice and identify any outstanding concerns. It was a real melding of people and resources."

Murray Swyripa
Vice President, Environmental Affairs

Whatever mine design DDMI chose would have to minimize environmental effects and fit the needs of local communities.

"We considered it vitally important that baseline information be assembled as early as possible to ensure the development of a credible record," recalls John Stephenson. "Preservation of water quality in the lakes and drainage systems was of paramount importance to the Dene and Inuit communities who had expressed their great sensitivity to sustaining the environmental quality of the land and water."

Understanding the Land and Water

The area studied included not only the area around the project site, but also the Lac de Gras watershed and beyond. DDMI hired independent contractors to sample, study and report on any effects the diamond mine might have on the environment.

DDMI also had a number of employees who were focused on environmental protection. Erik Madsen, a familiar Northerner who had worked in the environment field for the Government of Canada and several mining operations, was hired to be DDMI's Environmental Manager. He brought to DDMI not only his knowledge and contacts, but also a great respect for the land. Another key member brought on to the environmental team was Gord MacDonald, who managed the linkage between the consulting groups preparing the environmental assessment and the project engineers. "People always ask if it was difficult to get engineers to consider environmental concerns. What was different with Diavik was that the entire project team, starting with the President, understood the importance of the environment at this site and accepted the cost of the necessary precautions. That made my job relatively easy."

Murray Swyripa, an environmental scientist who had been extensively involved in environmental assessment work with the federal government, became DDMI's Vice President, Environmental Affairs: "From its very early stages, Diavik recognized that it must take the lead to ensure community involvement in environmental protection and socio-economic matters associated with our project. This approach was consistent with the policy directions of both levels of government as well as what we were hearing in the communities."

Traditional knowledge contributions to the baseline studies came from community residents and elders. This traditional knowledge supplemented other scientific studies that were conducted by the company. DDMI's studies measured and assessed fish and water, wildlife, vegetation and terrain, air quality, heritage resources and socio-economic conditions. Potential cumulative effects were also examined in respect to the establishment of the Diavik mine.

Community Consultation Process

In the early 1990s and throughout the succeeding years, representatives from the Diavik Project held regular meetings with community leaders and elders to inform and discuss all the project activities.

Darrell Beaulieu, the Chief Executive Officer of Deton'Cho Corporation and at the time, Chief of the Yellowknives Dene in N'dilo, remembers the project team's community approach. "Very early on, prior to the project's description, Diavik started communicating with the community

The company's plans for progressive reclamation, restoration of fish habitat and mine closure were discussed at community meetings. Murray Swyripa explains the steps that will be taken to avoid adverse environmental effects during the mine construction and operations.

In the spring, the crystal clear waters of Lac de Gras are among the purest found anywhere on earth.

leaders and the community itself," he said. "It was part of an education process to bring everybody up to speed, to explain what they were doing, how they were going to do it, when they were going to do it. So they've taken that time to explain things as best as they could."

When the project moved into advanced exploration in 1995, the Diavik Project team continued to nurture the consultation process especially with the neighbouring aboriginal communities affected by the mine – Wekweti, Gameti, Wha Ti, Rae/Edzo, Dettah, N'dilo, Lutsel'Ke, Kugluktuk, and the North Slave Métis Alliance – which were designated "primary impact communities." Yellowknife was also identified as a primary impact community, but because of its size and economy, it was placed in a different category.

To promote this consultation process, it was decided that meetings would be held in communities whenever possible. As is the local custom, meetings would often start and end with a prayer by the host community. The meetings were conducted in the language of the host community and English, with translation provided to facilitate mutual understanding. In accordance with tradition and culture, the views of elders were given prominence.

From the meetings held between 1994 and 1999, DDMI created a database of issues, concerns and questions. The company followed up on these

Community meeting in Rae

A Meeting in Dettah

Erik Madsen's first meeting after joining the DDMI team was in Dettah during the spring of 1996. About 15 people from the Yellowknives Dene attended, and representing Diavik along with Erik Madsen was Doug Willy. "Having lived in Yellowknife all my life, I personally knew a few of the younger members. Some of the elders I had seen around town," said Madsen. "But before Doug could get into some further details of the project, an elder named Judy Charlo spoke out in her native tongue. Judy talked for 45 minutes, telling us how the mines in the Yellowknife area had ruined the land, killing the trees. That her people could not eat the fish and drink the water anymore. That animals, especially caribou, did not come close to their community anymore. When she finished, Doug simply told her that she was no doubt right. The way mining was done in the past created some problems. But we wanted to hear their concerns so those ways would not return. We said that we wanted this future mine to be their mine too, and the only way that it would work is if we worked together to help design it. Looking back, I feel this first meeting in Dettah was the start of a positive relationship with the Yellowknives, where trust was reached and there was a definite will there to work together. When we came back for another meeting, Judy Charlo was the first to welcome us."

issues by taking action to incorporate them into the project. If the company could not incorporate them, it went back and explained why.

Doug Willy recalls, "When we went to these meetings, the point was not to tell anybody anything; the point was to listen to what the people had to say. The history of the mining industry in the North would often be raised. And then we would say: 'Okay, give us a chance.' We listened to them, they listened to us. They thought about what we had said and came back to us with their comments. We took them very seriously."

In addition to the numerous formal community meetings, the informal meetings were too numerous to count. "Sometimes we would get a phone call from someone who said, 'Hey. Let's meet for coffee.' So we did," remembers Davey. "And a lot of those meetings, we never even discussed diamonds, mines, or any other aspect of the project. We just talked about other things, things that had nothing to do with diamond mines but things that were important to the people in the communities. We had a goal to create relationships with the people in the communities, to get to know them better, and in the process, hopefully to become friends."

Doug Willy gets suggestions from Darrell Beaulieu and Jonas Sangris on moving forward.

DDMI will need to build water diversion dikes in order to access the diamond bearing ores beneath Lac de Gras. Mike Baker and Brian Saul explain the details of DDMI's mining and environmental plans to community members. In each community visited, valuable information and constructive suggestions were received.

"Rio Tinto believes that positive relationships with the communities that host our mining operations are essential. It takes time to lay the foundation upon which trust and mutual respect is based. I want to pay tribute to the people in the neighbouring communities who are working with the Diavik project team and who are making the time available to advance our shared interests."

Gordon Sage
Former Chairman of DDMI and
former Executive Director,
Industrial Minerals, Rio Tinto

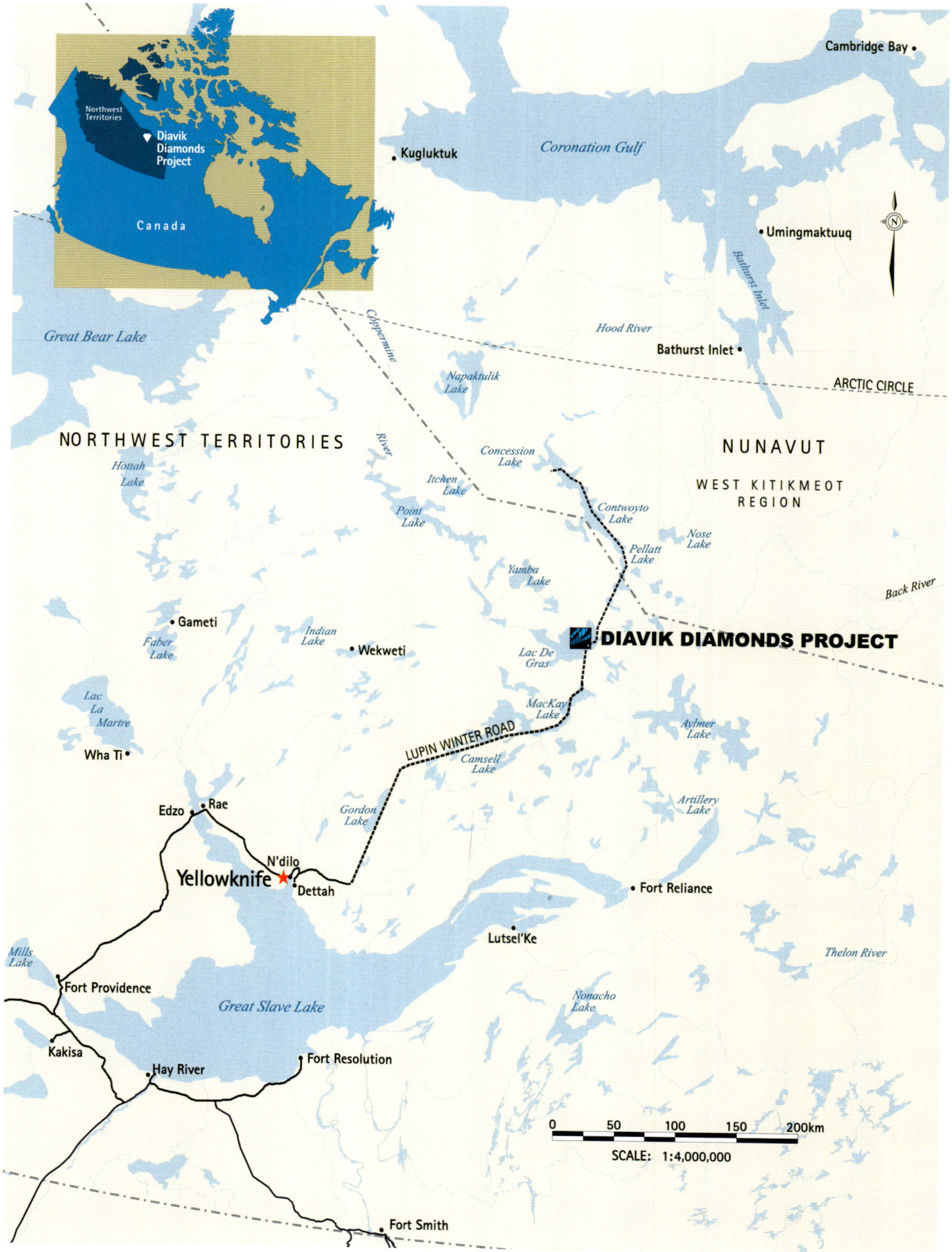

Northwest Territories
Diavik Diamonds Project
Canada
Cambridge Bay
Kugluktuk
Coronation Gulf
Umingmaktuuq
Bathurst Inlet
Hood River
Bathurst Inlet
Great Bear Lake
Coppermine River
Napaktulik Lake
ARCTIC CIRCLE
NORTHWEST TERRITORIES
NUNAVUT
WEST KITIKMEOT REGION
Hottah Lake
Concession Lake
Itchen Lake
Point Lake
Contwoyto Lake
Pellatt Lake
Nose Lake
Yamba Lake
Back River
Gameti
Faber Lake
Indian Lake
Wekweti
DIAVIK DIAMONDS PROJECT
Lac De Gras
MacKay Lake
Lac La Martre
Wha Ti
LUPIN WINTER ROAD
Camsell Lake
Aylmer Lake
Artillery Lake
Gordon Lake
Edzo
Rae
N'dilo
Yellowknife
Dettah
Fort Reliance
Lutsel'Ke
Mills Lake
Fort Providence
Thelon River
Great Slave Lake
Nonacho Lake
Kakisa
Hay River
Fort Resolution
0 50 100 150 200km
SCALE: 1:4,000,000
Fort Smith

CHAPTER 6

Bridges to the Future

"[Diavik] is a welcome development. It's a tremendous boost to our economy, there's no doubt about it. It's going to provide jobs, it's going to provide contracts to our communities."

The Honourable Stephen Kakfwi, Premier
Government of the Northwest Territories

Premier Kakfwi

The environmental approval process entered a more formal stage in early 1998. Under the leadership of the Government of Canada a steering committee was established to guide the environmental approval process. The steering committee, comprising local aboriginal communities and both levels of government, determined that the process would be as inclusive and transparent as possible.

Building on DDMI's earlier consultations, numerous opportunities were provided for aboriginal communities and the general public to participate and provide input into the project's design and development plans. Community groups and non-governmental organizations committed considerable resources to the review process. The federal government also dedicated significant resources, supporting the process with external technical expertise thereby adding to the technical rigour of the project review.

"The Participation Agreement is an excellent example of Kitikmeot Inuit and Diavik working cooperatively so the benefits from the mine go to the communities. As well, the Environment Agreement gives us confidence that the lands, wildlife, and water will be looked after for future generations."

Charlie Evalik
President, Kitikmeot Inuit Association

In early November 1999, the federal Minister of Environment gave official approval to move forward with permitting and licensing of the project.

The Minister concluded that with the implementation of all the mitigation measures identified, the proposed project was not likely to cause significant adverse environmental effects. The Minister emphasized the importance of continued meaningful involvement of aboriginal people in the monitoring of project activities over time and the implementation of a progressive reclamation plan for the site.

Formalizing Commitments

In parallel with the final stages of the environmental review of the project, the Government of the Northwest Territories and DDMI entered into discussions leading to the signing of the Socio-economic Monitoring Agreement (SEMA) for the project. The SEMA was the first leg of a three-part plan DDMI wished to put in place to encourage continuing community involvement in project activities following project approvals. "While at the time we were not too certain of the exact path events would take, we were hopeful that the Socio-economic and Environmental Agreements together with our proposed Participation Agreements with our five neighbouring aboriginal groups would provide a useful framework for continued community involvement" recalls Stephen Prest, who had become DDMI President in the spring of 1999.

Under SEMA, DDMI and the Government of the Northwest Territories commit to a number of undertakings with respect to employment, training, business development, capacity building, scholarships and community development. The SEMA also provides for an Advisory Board with representatives from aboriginal communities, the territorial government and the company to review and monitor the socio-economic impacts of the mine.

The Department of Indian Affairs and Northern Development (DIAND) began implementation of the environmental assessment recommendations shortly after the November 1999 approval, including establishment of an Environmental Agreement.

"The Environmental Agreement was groundbreaking. First, it's founded on community-based monitoring, in addition to the normal regu-

Joe Rabesca, Grand Chief of the Dogrib First Nation and Stephen Prest enter into a Participation Agreement between the Dogrib Treaty 11 Council and DDMI.

latory monitoring that takes place for a project like Diavik. And second, the aboriginal communities will appoint the majority of the Environmental Monitoring Advisory Board members. That's the first time that's happened in Canada," said Bob Turner, Vice President, North Slave Métis Alliance and Chair of the Diavik Environmental Monitoring Advisory Board.

The Environmental Agreement sets out the environmental monitoring and reporting requirements for the project; the framework for progressive reclamation of mine workings to achieve incremental abandonment in a manner consistent with sustainable development; mechanisms for DDMI's participation in a Regional Cumulative Effects Management Forum; and environmental security consistent with life-of-mine closure costs.

In addition to DDMI, the Parties to the Diavik Environmental Agreement include DIAND, the territorial Department of Resources, Wildlife and Economic Development, the Dogrib Treaty 11 Council, the Lutsel'Ke Dene First Nation, the Yellowknives Dene First Nation, the North Slave Métis Alliance and the Kitikmeot Inuit Association.

"Grand Chief (Joe) Rabesca has spoken today of this occasion as a signal of our collective desire to work together in the spirit of partnership and mutual respect. For Diavik, the challenge over the next twenty years will be to accomplish all that it has set out to do, to meet its commitments under the present participation and other agreements, thereby earning the continued respect of the Dogrib people."

Stephen Prest
President

On the day of the signing of the Participation Agreement, Lutsel'Ke elders, community leaders and DDMI staff joined in a signing ceremony and feast in the community. About 150 people sat down for a celebratory feast.

As the formal approval process and agreements were finalized, DDMI proceeded to formalize the commitments made to aboriginal communities in the form of individual Participation Agreements. These agreements detail the principles of mutual respect, active partnership, and long term commitment that guide the working relationship between DDMI and the aboriginal communities.

The first Participation Agreement was signed in March 2000, with the North Slave Métis Alliance, followed shortly thereafter with the Dogrib Treaty 11 Council in early April.

In October 2000, the Yellowknife Dene First Nation became the third aboriginal group to sign a Participation Agreement with DDMI. The Kitikmeot Inuit Association and the Lutsel'Ke Dene First Nation also signed Participation Agreements with DDMI in September 2001.

All the commitments made, agreements signed and policies implemented relate back to DDMI's basic philosophy of encouraging northern

Yellowknives Dene First Nation Chiefs, Richard Edjericon and Peter Liske formalize the Participation Agreement with Stephen Prest and Hubie van Dalsen from DDMI.

"What we would hope could be achieved, say in 50 years time, is that when people talk about Diavik, they'll remember it fondly. They'll remember Diavik as a good neighbour, somebody who was there to help, and made positive contributions to the society we lived in. That, to me, is the legacy that we at Diavik must be striving for."

Stephen Prest
President

communities to participate in the project. These reinforce the company's commitment to conduct all its activities in a manner consistent with sustainable development. By integrating the economic, environmental and community interests, the project will create enduring community benefits.

"This company said it will continue to have a community presence, to build and make sure the communities have an in, so anyone from the community knows they've got a door to the project, and the fundamental commitment is that the door will always be open," said Eric Christensen. "Diavik will continue to work closely with the communities throughout the life of the mine. Even 10 to 15 years down the line, Diavik will continue to subscribe to the principles of active partnership, mutual respect and long-term commitment, living up to the agreements we have signed."

A handshake between Clem Paul, President of the North Slave Métis Alliance, and Stephen Prest seals the Participation Agreement they had just signed.

The Go Ahead

"This is going to be very big business for Canada in general, and in particular, for the North, of course. There is certainly a romance attached to diamonds being in this kind of location."

Bob Gannicott
President and Chief Executive Officer,
Aber Diamond Corporation

By August 2000, the regulatory framework for the project was in place. In the following months, the Mackenzie Valley Land and Water Board reviewed and issued the specific approvals that authorized DDMI to proceed with construction of the mine facilities. In December 2000, DDMI and Aber Diamond Corporation announced the decision to proceed with the development of the mine.

This decision cleared the way for DDMI to begin the process of bringing into effect its numerous agreements – Environmental, Socio-Economic and Participation Agreements – and to begin with the construction of the mine and support facilities. Canada's second diamond mine was on track to start up for operations in the first half of 2003.

Completing the initial phases of the project took commitment, dedication and hard work. Taking it further into construction, operation and finally, closure of the mine will require even more commitment and dedication from the company and its people. It is this level of dedication that will ensure that DDMI will protect the environment while creating enduring benefits for people of the North.

"About the same time Diavik was going through the approvals process, the federal government was moving forward with an action plan to work together with aboriginal people on key priorities that would result in strong, healthy communities and an improved quality of life for aboriginal peoples.

The Diavik project was quite timely as the new policy direction focused on building partnerships with aboriginal people and encouraging the private sector to play a role in building northern community capacity. The involvement of the new Mackenzie Valley Land and Water Board in the implementation of Diavik's water licence has also been key in giving Northerners a greater say over land, water and environmental matters."

Lorne Tricoteux
Associate Regional Director, DIAND

CHAPTER 7

Let the Work Begin

"Times have changed. Mining companies like yours now come and talk to us about their plans and ask for help, not like before. As long as the land, the water and the animals are taken care of, the mine should be built. We need jobs for our young people to keep them out of trouble. I want to see that mine built before I die."

Michel Paper, 85 year old Elder from Dettah

Above: A kimberlite sample

Left: Construction of permanent fuel tanks at the Diavik site.

The Diavik Project is one of the largest capital investments ever undertaken in the history of Canadian mining. The magnitude of the capital required for the project outstripped the funds available from the Canadian private sector for this type of mining venture. As a result, in parallel with DDMI's environmental and community activities, the management of DDMI and Aber worked independently behind the scenes to raise their portion of the capital necessary to finance the project.

In DDMI's case, its share of funding for the project was raised on international markets through its parent company, Rio Tinto. As a publicly traded Canadian company, Aber turned to the North American investment and banking communities and other capital sources.

Dogribs' Blessing of the Diavik Diamonds Project

On a blustery spring day in 2000, three Dogrib elders – Joe Migwi, Eddy Weyallon and John Dryneck along with translator James Rabesca – and DDMI representatives – Stephen Prest, Doreen Sangris, Tom Hoefer, John Tees, Doug Willy and Len McHale – conducted a traditional Dene spiritual ceremony with drums, tobacco and a prayer at the main camp of the mine.

The ceremony demonstrated DDMI's commitment to join aboriginal communities in respecting their traditions. The ceremony acknowledged the presence of the Creator, and offered prayers for the safe conduct of work and thanks for the land, water, air and animals. The Dogrib elders presented Stephen Prest with a prayer drum so he could participate in the ceremony.

This cooperative gesture between the company and the Dogrib leadership under Grand Chief Joe Rabesca was, according to the Dogribs, the first time a mining company had made the effort to respectfully embrace the spirituality of the Dogribs and invite them to conduct such a ceremony.

The four Diavik pipes contain reserves of 26 million tonnes of kimberlite. An average annual production rate of six million carats is expected. But these estimates may vary as the pipes develop and the mining matures from open pit to underground. Based on these estimates, the life of the mine is expected to be 20 years.

For the first 10 years of operation, the Diavik Project will contribute in excess of $2 billion in direct tax and royalty revenues to various levels of government. During peak production, Diavik will provide about five per cent of the world's diamond supply.

Designed for the North

Beginning in 1996, DDMI conducted engineering design and project feasibility studies which underpin all aspects of the project. DDMI assembled a team with years of experience in large construction projects. Many members of the project team were drawn from DDMI's parent company, Rio Tinto and its subsidiaries. Rio Tinto is one of the largest mining houses in the world with projects on almost every continent. DDMI also engaged Canadians who had the knowledge, skill and experience needed to manage the design and construction of the mine. Nishi-Khon/SNC-Lavalin (NKSL), a Dogrib joint venture company, with significant experience in development of arctic projects, joined the Diavik team in 1997 and were key to successfully developing the basic project designs to suit the northern environment. In addition to NKSL, many other northern companies made valuable contributions to the design.

"Diavik is a fascinating project – it is a very satisfying engineering challenge. The water diversion dike is the backbone of the project. Without it, we could not safely access the kimberlite pipes below the lake. Think about it. We are building a very long dike with its base deep under water. The consultations between the company and the communities – seeking their input into the design of the mine including the water diversion dike – have been key. That input has definitely resulted in a better project design," said John Wonnacott, Deputy Project Manager.

DDMI established its project office in Calgary in 1996 to facilitate bringing together the people who would assemble all the information needed to be incorporated into the project joint venture feasibility study. Former Vice President, Finance,

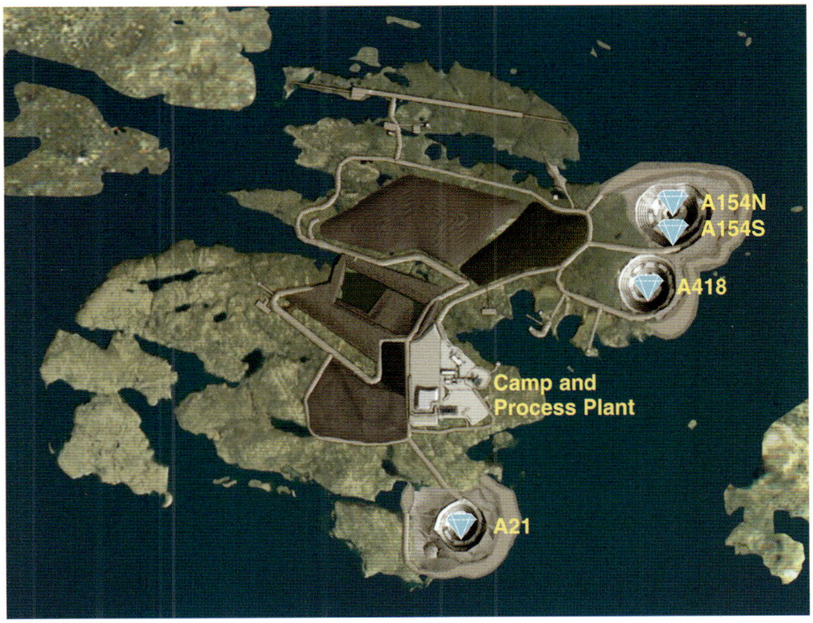

Conceptual illustration of Diavik Project site at full development

"Project management in the North has got its own uniqueness and challenges. Special care for the environment, incorporation of local people and accessibility to the site during the construction phase are only some of them. It tends to be an exercise in logistics as engineering, procurement and construction activities are controlled by climate and remoteness."

Philip du Toit
Vice President
Project Management

Hubie van Dalsen recalls those days, "There was a lot of traveling back and forth between DDMI's offices in Calgary and Yellowknife by a steady stream of geotechnical and other engineers, mine planners, administration people and environmental specialists whose task it was to collect the technical data on the project and to participate in community meetings."

A secure wing was built in the Calgary office to house the Diavik diamond samples from time to time. Jim Sharp, former Vice President, Sales and Marketing, recounts, "The 21,000 carats from the bulk sample from A154 and A418 were of such quality to justify moving forward with the project. When shown the Diavik samples, those knowledgeable in the trade were genuinely enthusiastic about this potential new addition to the world's diamond supply."

Diane Looffen examines rough diamonds from the Diavik Project.

Surveying site lines at dusk.

The Lupin Winter Road

Since 1982, the Lupin Mine winter road, built by Echo Bay Mines, has been a major artery for shipping supplies and materials to isolated projects in the North. Shipping by air is extremely expensive, so the winter road, which winds its way from the end of Ingraham Trail near Yellowknife to the Lupin Mine at the north end of Contwoyto Lake in Nunavut, plays a major role in any northern construction project in the area. The winter road is constructed every year along the same route, 75 per cent of it across the ice of northern lakes. Countless numbers of trucks have safely hauled thousands of loads during the short, 10-week ice road season that usually runs from late January to early April. Trips are timed and carefully monitored to ensure the safety of the drivers and to minimize damage to the ice. About 4,000 truckloads made the 416-kilometre trip to Diavik in 2001. Approximately 1,800 loads will be needed during each year of operation. Diesel fuel to power generators and mobile equipment will be the single largest commodity being hauled. Although mainly used as a mine supply route, the winter road is also used by the general public.

Building the Mine

In spring 2000, the rush was on. DDMI had to ship several thousand tonnes of materials, fuel and supplies – more than a thousand truckloads – up the winter road. Trucks rolled up the glistening blue ice ribbon that traverses lakes and portages for more than 300 of the over 400 kilometre distance between Yellowknife and the mine site. Water, transformed by nature's frigid blasts into rock hard ice measuring up to two metres in depth, serves man and caribou alike throughout the long arctic winter. The Bathurst caribou travel great distances during their seasonal trek north to spring calving grounds. Now, just as the aboriginal peoples before them, DDMI follows their path north in an annual ritual of replenishment, not this time for the hunt, but instead to draw upon nature's bounty from deep within the land.

DDMI's approach to development of the Diavik Project has always recognized that the land and water are inextricably linked in Canada's North and must be handled with care and respect in their pursuit of diamonds from the depths below. Diavik requires dike structures to temporarily divert the waters of Lac de Gras from the future mine workings. Construction of the first water diversion dike, around the A154 kimberlite pipes, will use rock quarried from the island and placed in the lake to extend the land in a great four kilometre looping arc from the shoreline of East Island.

Much preparatory work has been started. Support facilities, fuel storage, accommodations, and an airstrip are required. DDMI's building plans, all reviewed by the communities and elders, include special features to minimize disruptions to the land, fish habitat and the caribou who traverse the island. Because the project involved building the mine on an island in the middle of a pristine arctic lake, there was little room for error in predicting what the water quality impacts would be. Mike Baker, Environmental Specialist, commented "Very early in the process, DDMI committed significant resources to its baseline geochemistry program. This allowed DDMI to undertake a thorough evaluation of the interactions between the water and the rock once it is brought to the surface and plan the necessary water treatment to protect Lac de Gras."

Doug Laverdiere, Gem Steel, working with Robert Lafferty of Rae, one of the first graduates of the community-based training program conducted in partnership with Aurora College and the governments of the Northwest Territories and Canada.

"The participation of community members including elders and leaders in selecting and monitoring Diavik's community based training programs allowed us to develop among our trainees the skills base needed for Diavik's construction and production activities."

Glenn Zelinski
Manager, Construction Training

Caribou

To the traditional peoples of the North, the caribou is integral to their lives and cultures. Inuit call the caribou "tuktu" or "tuktuk", Chipewyan people say "et-then", in Dogrib the word is "ekwo", and to the Slavey the caribou are "ekwe".

The caribou is a member of the deer family. There are three types of caribou in Canada: woodland, barren-ground, and Peary.

Barren-ground caribou make up the majority of caribou in Canada and are the mainstay of many northern communities. There are approximately 1.2 million barren-ground caribou in Canada, most living in five large herds known as the Porcupine Herd, Bluenose Herd, Bathurst Herd, Beverly Herd, and Kaminuriak Herd. These herds migrate seasonally from the tundra to the sparsely treed northern coniferous forests. Some of the 350,000 strong Bathurst Herd migrate through the Lac de Gras region.

Closure and Reclamation

DDMI is employing the concept of progressive reclamation, which simply means, the company is cleaning up as it goes. "If you know in advance how you're going to close the project, then you can develop a plan to implement the proper closure dynamics, particularly if there are areas that might be sensitive," said Stephen Prest. "Progressive reclamation was adopted as part of our project submission. Because of that, we, together with government and our aboriginal communities, were able to recognize progressive reclamation in the Environmental Agreement reached between us. As far as I know, that's the first time that's been done in Canada."

How to deal with the two large rock piles left behind by 20 years of mining activity was an issue in designing for closure. Since caribou sometimes use part of East Island for their annual migration, it was important not to create a barrier for the animals as they travel their ancient paths.

The traditional knowledge collected as part of DDMI's community consultation and mine design, showed that caribou naturally seek higher ground because breezes in open areas provide relief from the onslaught of insects. Working from the suggestions by the elders, DDMI engineers have incorporated gradual ramps on every side of the rock piles, so caribou can easily climb them. As more and more rock is piled on the island, these ramps will be continuously refined so that the caribou can continue to use the land as they have for the past millennia.

"The people of the North take pride in their pristine environment. Their time frame is not measured by a 20-year project, but rather by several generations. During the pre-feasibility phase of Diavik, the engineering design that was developed placed great emphasis on the northern people's perspective. With this in mind, we started with closure of the project and worked backwards to start-up, placing emphasis on a several generation view of the land."

Jeff Tygesen
Former Manager, Mine Development

April Desjarlais taking sediment samples.

Creating Aquatic Habitat through Reclamation

The fact that the mine pits are located on the edge of Lac de Gras presents an opportunity to reclaim the pit areas to become, once again, part of the lake. As part of this process, portions of the dikes constructed to permit mining in the first place, will be removed sometime in the future to allow free-flow of lake waters over the areas where the company has mined.

As part of DDMI's closure plan, fish habitat and artificial islands will be created in areas where the dikes once stood. These islands and shoals, once created, will render these shorelines an ideal spring habitat for migrating shorebirds.

"As a result of discussions with local communities, the project closure plans changed through numerous meetings," said former Manager, Mine Development, Jeff Tygesen. "For example, the water retention dikes were reduced in size. The shapes of the dikes were modified to enhance fish spawning areas. The location and shape of the granite stockpiles were modified to enhance caribou migration and promote diverse vegetation. Our consideration of the northern people's view and their time frame, their perspective on future generations, ultimately enhanced the final design of the Diavik Project. A great deal of thanks goes to the northern communities for contributing their time and ideas to the Diavik Project."

Above: Red-throated loon

Left: Camilia Zoe Chocolate is a member of DDMI's environmental team responsible for ensuring the land, water and wildlife are taken care of.

CHAPTER 8

At the End of the Day

"The elders have instructed Diavik and the government to take of care of our land, the water and the caribou."

Rachel Crapeau, Yellowknives Dene First Nation
Yellowknife Public Technical Sessions – March 1, 1999

Above: Dwarf fireweed

Left: The annual migration of the Bathurst caribou herd is critical to the lives of northern aboriginal peoples.

A Final Word ...

From the beginning, Diavik has focused on being a northern mining company, one that would become part of the community of the North and keep a northern perspective throughout its lifetime.

Community consultation was, and still is, a fundamental part of DDMI's corporate philosophy. The mine is designed to fit the North. Aboriginal communities play major roles as employees, as business partners, and as guides on how to thrive in the North while respecting the waters and the land.

Increasingly, mining companies are expected to contribute to the creation of a better society to achieve the aims of sustainable development. The people building Diavik, and those who will run it, hope these things will remain long after the mine is gone – a permanent living legacy of stronger, healthier communities. Diavik is an opportunity to show that mining can be environmentally and socially responsible, creating a model for those who follow. For Diavik the path is clear, the course is set – to be a contributor to the sustainable development of the North.

President Stephen Prest on the subject: "If one has successfully conducted one's affairs in these matters, that is, developed the mine, operated it, used these activities for social and economic good in the host community, closed the mine and restored the land to its pre-mining uses and beyond, then all of us associated with the Diavik Project, past and present, and all those that follow will be able to take immense pride in our contributions to sustainable development in Canada's North."

Right: Conceptual illustration of the Diavik site on East Island following reclamation

Far right: Arctic cotton

Epilogue

And so, the first chapters of the Diavik Diamonds Project have drawn to a close but the journey continues as a river through time. The initial project period is not unlike the swift currents of the northern rivers at their headwaters. Replenished each spring, cascading over falls and rapids with but brief periods of still waters, running nonetheless true to their course.

Along the way the river encounters the great boulders of the rapids, the sands of the lakes, the silts of the barren land marshes, all contributing in their own way to its character as it ebbs and flows downstream. Many hundreds of people, if not thousands have contributed to the foundation of Diavik, also each in their own way. Thousands upon thousands of interactions, ideas and visions commingled over the years, tumbled together as the silts, sands and boulders of the great northern rivers, gathering strength on its downstream path – this is how Diavik has come into being.

In the pages that follow, we are proud to acknowledge many of the men and women who have contributed to the onward course of the Diavik Project. Thanks to their dedication we have the Diavik story to tell.

Stephen Prest
President
Diavik Diamond Mines Inc.

For centuries,
people of the North
have used the
resources wisely.
... Diavik is continuing
this tradition.

Alphonse Quitte

Diavik Diamond Mines Inc. Team

Sarah Archer
Doug Ashbury
Greg Bailey-Hopf
Bronwyn Baker
Dave Baker
Mike Baker
Darren Barlow
Serge Benoit
Darryl E. Bohnet
Rick Bonner
Doug Bryan
Jennifer L. Burgess
Dina Lori Cardinal
Darcy Chattell
Eric Christensen
Michael Cooper
Harley Crowshoe
Bob Dawe
John DeBoer
April Desjarlais
Philip du Toit
Arnold Enge
Tim Evans
Kelly Fahl
Alestine Ferdinand
Adrienne Forest
Helen Francois
Valerie Frise
Jeff Giesbrecht
Cindy Gilday
Steve Goudie
Kimberley Guay
Ron Hampton
Nancy J. Hanks
Barry Haydon
Peter Heine
Denton Henkelman
Laurie Hilkewich
Tom Hoefer
Doug Ivey
Evelyn P. Kublu-Hill
Brenda J. Kuzyk
Benoit Lacasse
Pierre Leblanc
Richard Lock
Gordon MacDonald
John D. Mackay
Nicole Mackay
Bertha Mackenzie
Grace Mackenzie
Gabrielle Mackenzie-Scott
Kimberly MacLeod
Elizabeth MacPherson
Erik Madsen
Mariah Maring
Len McHale
William McIntosh
Wade McNenly
Belinda Mercredi
Jonathan T. Metcalfe
Brian Mitchell
Lawrence Monks
Wilma Nicholson
Richard Olive
Laura Ettagiak Orchard
Durward Pankow
David Passfield
Sandy Penkala
Louise Picard-Hampton
Stephen Prest
Susan Reddick
Bea Rodh
John Rogers
Martin Roseke
Tracey-Lyn Roseke
Debbi Ross
Greg Rutherford
Harry Ryans
Doreen Sangris
Brian Saul
Jim Sharp
Wendy Sheridan
Marion Storm
Nick Strus
Michael Sullivan
Murray Swyripa
John Tees
Garry Tkachuk
Jocelyn T. Tonge
Tim Townsend
Dale Tweed
Elizabeth van Heerden
Schalk van Heerden
Leah Von Hagen
Kimberly Williams
Sean Willy
James Kam Piu Wong
John Wonnacott
Cheryl Wray
Scott Wytrychowski
Calvin G. Yip
Francis Zacharias
Glenn Zelinski
Henry Zoe
Camilia Zoe Chocolate

Blueberries in late summer

Past Diavik Diamond Mines Inc. and Kennecott Canada Inc. Contributors

Lee Able
Larry Adams
Mark Aitkenhead
Melaw Antoine
James Arden
Ken Armstrong
Peter Arrowmaker
Kim Auge
William Auger
Joseph Bailey
Alfred Baillargeon
Jonas Baillargeon
Faye Ballas
Frank Basil
Henry Basil
Marcel Basil
Angus Beaulieu
Herbert Beaulieu
Mary Betsina
Norman Betsina
Johanne Lee Black
Alison Blackduck
Francis Blackduck
Myranda Bolstad
Jerome Bourchier
Ian Boxall
Joe Boysis
Nick Brajkovich
Shannon Brockman
Kathleen Burris
Melissa Casaway
Iris Catholique
Larry Catholique
Corey Champion
Owen Chan
Karl Chappel
Marcel Chartrand
Regan Chernish
Julia Christensen
Leslie Coe
John Collier
Roland Conrad
Valerie Conrad
William Davis
Andy Davy
Rod Davey
Matthew Dawe
Myra Delorme
Kenny Desjarlais
Stan Desjarlais
Wally Desjarlais
Rhonda Desnomie
Dick Dustal
Hunter Dickson
Noel Doctor
Buddy Doyle
Bobby Drybones
Gary Drybones
Tony Dryneck
Scott Dunbar
Vernon Edjericon
Darka Edwards
Leon Ekendia
Issac Embodo, Jr.
Jerry Enzoe
Randy Evans
Casey Eyakfwo
Wade Faryna
Ron Fatt
Zane Fennell
Mark Fields
Eleanor Firth
Johnny Fish
Niall Fitzgerald
Michael Francois
Barry Franklin
Michael Fraser
Jeremy Frizzell
Sesino Gacayan
Christopher George
Norman Gerwing
David Giroux
George Goulet
Ian Graham
Waylon Greenland
Shawn Greensides
Marlene Grooms
Vincent Halushka
Royden Hamilton
Andrea Harrison
Falck Hendrick
Gordon Hosick
Lena Hughes
Joe Inman
Misty Ireland
Joanne Irving
Steven Itsi
Joel Jansen
Denise Johnson

Adeline Judas
Roy Judas
Richard Kailek
Dave Kelsch
Erin Kimball
Kevin Kivi
Earl Kock
Colin Koe
Fred Kotchilia
Krista Kreger
Ken Kuchling
George Lafferty
Jackson Lafferty
Jane Lafferty
Natasha Landry
Mark Leblanc
Shawyne Lemouel
John Lennie
Grzegorz Lickiewicz
Alfred Liske
Alice Liske
Joseph Liske
Diane Looffen
Chris Lord
Casey Lunn
Ronna MacCara
Rob MacIntosh
Paul Mackenzie
Calum MacPherson
Gaeleen MacPherson
Murdo MacPherson
Scott MacPherson
Daniel Mantla

Marnie Marchuk
Anthony Margarite
George Martin
Angela Masongsong
Mark McDowell
Cynthia Meinhardt
Ruth Menzies
Peter Mercredi
Debbie Michon Weir
Cheryl Moore
Kathy Moore
Jennifer Morin
Michael Morin
Paul Mounsey
Ted Muraro
Leon Nasken
Della Nault
Ian Neill
Steven Nitah
Brie O'Keefe
Matthew O'Keefe
Ollie Olsen
Joslyn Oosenburg
Andrew Oxford
Sherry May Palichuk
Jessica Patterson
Tom Patton
David Paul
Chris Petrovic
Lucien Puech
Frederick Rabesca
Peter Ravenscroft
Tania Rebalkin

Chuck Richardson
Paul Rowe
Winston Rowsell
Richard Rusden
Joanne Saftich-Cervin
Keith Sangris
Peter Sangris
Lori Schumann
Amos Scott
Mike Senn
Jim Sharp
Walter Shupe
James E. Siddle
Jessica Simpson
Bob Sinclair
Colin Smith
Jody Smith
Karl Smith
Shelley Stafford
John Stephenson
Tami Sutley
Leilagh Tate
George Tatsiechele
Derrold Taylor
Abemelek Tessema
Alfredo Tiburcio
Peter Tinqui
Lena Tiriraniaq
Brad Tricoteux
Mustafa Tuzun
Jeff Tygesen
Chinta Unka
Rueben Unka

Hubrecht van Dalsen
Rob Van Egmount
Pat Varas
Robert Villeneuve
Melody Voykin
Mike Waldman
Michelle Ward
Peter John Washie
Lloyd Wedzin
Richard Weishaupt
Peter Wellin
Phoebe Wetrade
Richard Weyallon
Sean Whiteford
Byron Whitford
John Wifladt
Brad Williams
Doug Willy
Loch Willy
Ashley Woodward
Roy Yipp
Charlie Zoe

Aber Directors and Employees

CURRENT

Directors

James N. Fernandez
Robert A. Gannicott
John C. Lamacraft
The Hon. Donald S. Macdonald, P.C., C.C.
John H. Parker, O.C.
J. Roger B. Phillimore
D. Grenville Thomas
Eira M. Thomas

Employees

Andrew B. Adams
Susan Annett
Michael A. Ballantyne
Elena Baltzis
Alia Bensimon
Joan A. Butt
Kuljit Chahal
Rod Chisholm
S. Caroline Glasbey
Farhad Keshavarz
Matthew L. Manson
Magdalina Quinn
Alex S. Speirs
Kaarina Venalainen

PAST

Directors

Earl E. Curry
Jordan C. Ethans
Kenneth G. Hanna

Employees

Ken Armstrong
Lee Barker
Alan J. Bayless
J. Frank Bradley
Shannon Brockman
Wanda Costuros
Jody Harris
Sarah Hasek
Robert Hindson
Robin Hopkins
R. Michael Jones
Leni Keough
Cathy Larkin
Casey Lunn
Zara Mitchell
Bill Mosher
Brenda Novak
Nosh Vellani

Caribou on the move

Photography

DIAVIK COLLECTION
Tim Atherton
Page Burt
Jiri Hermann
Paul Nicklen
Diavik staff

OTHERS

pi	Wayne Lynch
p3	Janice Phillips
p5	Wayne Lynch
p7	Wayne Lynch
p8	courtesy NWT Archives
p9	courtesy NWT Archives
p13	courtesy NWT Archives
p23	Terry Parker, courtesy NWT Tourism
p53	Wayne Lynch

The Diavik Project North Camp